Grade 5 · Unit 4

Inspire
Science

Earth and Space Patterns

McGraw Hill Education

Mheducation.com/prek-12

STEM McGraw-Hill is committed to providing instructional materials in Science, Technology, Engineering, and Mathematics (STEM) that give all students a solid foundation, one that prepares them for college and careers in the 21st century.

Send all inquiries to:
McGraw-Hill Education
8787 Orion Place
Columbus, OH 43240

ISBN: 978-0-07-699680-3
MHID: 0-07-699680-8

Printed in the United States of America.

9 10 11 LWI 26 25 24 23 22 21

Table of Contents
Unit 4: Earth and Space Patterns

Earth's Patterns and Movements

Encounter the Phenomenon 2

Lesson 1: **The Role of Gravity** 5

  Hands On Crater Model 8

  Hands On Falling Water 16

Lesson 2: **Earth's Motion** 23

  Hands On Shadow Measurements 26

  Simulation Earth's Movements 30

  Data Analysis Three Cities 36

STEM Module Project Planning 43

STEM Module Project **Design a Planetarium Model** 46

Module Wrap-up 47

Earth and Space

Encounter the Phenomenon 48

Lesson 1: **Earth's Place in Space** 51

  Hands On Model of the Sun, Earth, and Stars 54

  Hands On Model the Solar System 61

Lesson 2: **Stars and Their Patterns** 67

  Hands On Star Brightness 70

  Simulation The Night Sky 78

STEM Module Project Planning 85

STEM Module Project **Model a Constellation** 87

Module Wrap-up 89

Earth's Patterns and Movement

ENCOUNTER
THE PHENOMENON

Why does the shape of the Moon appear to change throughout the month?

Moon Phases

GO ONLINE

Check out *Moon Phases* to see the phenomenon in action.

Talk About It

Look at the photo and explore the digital activity of the Moon's phases. What questions do you have about the phenomenon? Talk about them with a partner.

Did You Know?

We always see the same side of the Moon from our view on Earth. Only 59 percent of the Moon's surface can be visible during its journey around Earth.

Design a Planetarium Model

You are an astronaut hired to collect data about Earth's movement. At the end of this module, you and a team of fellow astronauts will develop a model that will turn your classroom into a planetarium. You will have the opportunity to design a model of changes in the length and direction of shadows, the cycle of day and night, moon phases, or seasonal changes.

Lesson 1
The Role of Gravity

Lesson 2
Earth's Motion

PRIMARY SOURCE

Astronauts train for years before they go on space missions. They study physical science, space science, and even go through scuba and medical training. Most astronauts learn foreign languages so they can communicate with astronauts from other countries. Astronaut Mae Jemison, pictured above, learned to speak Russian, Swahili, and Japanese.

STEM Module Project

Plan and Complete the Science Challenge Use your knowledge of Earth patterns and movements to build your planetarium model.

Earth's Gravity

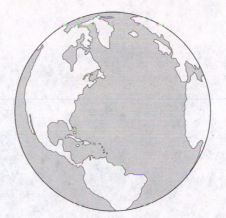

Four friends were talking about gravity. They each had a different idea. This is what they said:

Angie: I think Earth's gravity only pulls on things that are close to Earth's surface.

Tamara: I think Earth's gravity only pulls on things that are close to Earth and in Earth's atmosphere.

Carlos: I think Earth's gravity pulls on things close to Earth, including some things in space.

Mi-Ling: I think Earth's gravity pulls everything in the solar system toward it.

Who do you agree with most? _____

Explain why you agree.

_____

_____

_____

_____

_____

You will revisit the Page Keeley Science Probe later in the lesson.

The Role of Gravity

ENCOUNTER
THE PHENOMENON

What caused this crater to form?

GO ONLINE

Check out *Barringer Crater* to see the phenomenon in action.

Barringer Crater

Look at the photo of the Barringer Crater located in Winslow, Arizona. What questions do you have about the phenomenon? Record or illustrate your thoughts below.

Did You Know?

Barringer Crater in Arizona is the largest impact crater in the United States. It is almost 1.6 kilometers (1 mile) across.

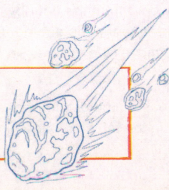

INQUIRY ACTIVITY

Hands On

Crater Model

Think about the crater you just saw. What causes the crater to form? You will investigate this phenomenon.

Make a Prediction How does the size of an object affect the size of the crater that it forms?

_____

_____

_____

Carry Out an Investigation

BE CAREFUL Wear safety goggles to protect your eyes.

1. Cover the floor with newspaper. Place the pan on the newspaper.

2. Fill the pan with flour so it is about two centimeters (cm) deep. Smooth out the flour in a flat layer.

3. Mold the clay into 3 different size balls.

4. **Record Data** Measure the diameter of each ball. Record the diameter of each ball in the data table on page nine.

5. Drop the clay balls from the same height into different parts of the pan.

6. Remove the clay balls from the flour by pressing a small piece of clay into the top and gently lifting up. Be careful not to move side to side. Set the clay balls off to the side.

7. **Record Data** Measure the diameter of each crater. Record the data in the table on page nine.

Copyright © McGraw-Hill Education (2)McGraw-Hill Education, (4)Michael Scott/McGraw-Hill Education, (5 6)Jacques Cornell/McGraw-Hill Education, (others)Ken Cavanagh/McGraw-Hill Education

Materials

safety goggles

newspaper

shallow aluminum pan

flour

ruler

modeling clay

Diameter of Clay Ball	Diameter of Crater

Communicate Information

8. How does the size of the craters compare to the size of the clay balls?

_____

_____

_____

9. Did your results support your prediction? Explain.

_____

_____

_____

10. How do you think this model represents how craters form on the surface of the Moon or planets?

_____

_____

_____

_____

INQUIRY ACTIVITY

11. Based on your data and observations, how would the size of a meteor affect the crater size left on a planet or moon's surface?

_____

_____

_____

_____

_____

_____

_____

_____

12. How do the results of this investigation model the role of gravity on Earth's surface?

_____

_____

_____

_____

_____

_____

_____

MAKE YOUR CLAIM

What force causes objects to fall to Earth's surface?

Make your claim. Use your investigation.

CLAIM

_____ affects how objects fall towards the center of Earth.

Cite evidence from the activity.

EVIDENCE

The investigation showed that _____.

Discuss your reasoning as a class. Tell about your discussion.

REASONING

The evidence supports the claim because _____.

You will revisit this claim throughout the lesson.

Look for these
words as you read:

gravity

meteor

meteorite

tides

The Pull of Earth's Gravity

Gravity is a force of attraction, or pull, between any two objects. The Barringer Crater is the result of gravity pulling a meteor to Earth's surface. The strength of gravity is affected by the total mass of the objects and the distance between them. The pull of gravity decreases when the total mass of the two objects decreases or they are further apart.

On Earth, gravitational pull is the attraction between an object and Earth. No matter the location on Earth's Surface, gravity pulls objects on all sides down toward the center. Every object that has mass experiences a gravitational pull. When an object is dropped from a certain height on Earth, it rushes downward due to gravity.

These skydivers have mass. Gravity is pulling them down toward Earth.

The Moon, Earth's closest neighbor, is greatly affected by Earth's gravity. Moons orbit planets for the same reason that planets orbit the Sun—because of gravitational attraction.

GO ONLINE Watch the video *Tides* to help you understand more about how the Moon affects oceans.

The Moon has less mass than Earth, so the Moon's gravitational pull is weaker. In fact, the Moon's gravity is about one sixth of Earth's gravity. Think about how high you can throw a ball on Earth. On the Moon, you could throw it about six times higher because the force of gravity is not as strong.

Earth's gravity affects the Moon. The Moon's gravity also affects Earth. The Moon's gravitational force causes Earth's **tides**, or the regular rise and fall of water along the shore. Earth's water bulges on the Moon-facing side of Earth. A bulge also forms on the side facing away from the Moon. The water level rises where the bulge is and the level lowers where it is not. This bulge of water causes changing tides as the Moon travels around the Earth.

MATH Connection The Moon's gravity is $\frac{1}{6}$ that of Earth. If an astronaut weighs 79 kilograms (175 pounds) and his space suit weighs 50 kilograms (110 pounds), how much would the astronaut and his suit weigh on the Moon?

PRIMARY SOURCE

In this photo, Astronaut Buzz Aldrin sets up an experiment on the Moon. Astronauts study the effects of gravity on Earth as well as in space.

Meteors and Meteorites

Objects other than the Sun and planets are found in our solar system. Sometimes, Earth's gravity will pull these objects into Earth's atmosphere.

Meteors You may have heard meteors be called shooting stars but a meteor is not a star at all. A <mark>meteor</mark> is a space rock that enters Earth's atmosphere. It appears as a bright streak in the sky. If a meteor does not break apart and burn up in the atmosphere, it can hit Earth's surface.

Meteorites A meteor that strikes Earth's surface is called a <mark>meteorite</mark>. Many places on Earth, like the Barringer Crater on page six, show evidence of meteorite impacts.

PRIMARY SOURCE

SEA OF SERENITY

SEA OF NECTAR

MARSH OF SLEEP

SEA OF FERTILITY

SEA OF CRISES

SEA OF WAVES

BORDER SEA

SMYTH'S SEA

This is a map of the Moon's surface made by NASA. It was used to help astronauts prepare for the *Apollo 10* mission. As you can see, there are many craters. These craters were caused by meteoroids hitting the Moon's surface. *Meteoroids* are small rocky objects that orbit the Sun.

Read the Investigator article *Gravitational Waves*.

WRITING ›Connection Write a paragraph about how the findings of Einstein and other scientists have been important in understanding how gravity works. How have the ideas of scientists in the past affected our use of technology in the study of gravitational waves? Answer the question using the graphic organizer below. Use a separate piece of paper to write your response in paragraph form.

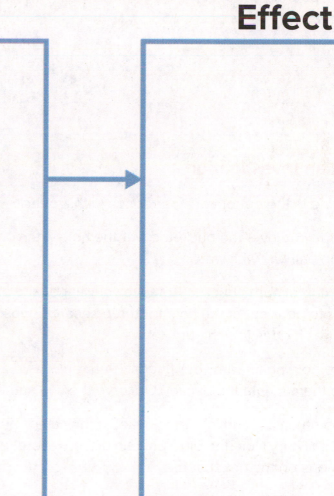

Cause

Effect

INQUIRY ACTIVITY

Falling Water

When you pour water out of a container, the water is pulled toward the ground.

Make a Prediction What will happen if I drop a container while water is pouring out of it?

_____

_____

_____

_____

Materials

2 paper cups

pencil

water

bucket

Carry Out an Investigation

1. Poke a hole in the bottom of a paper cup with a pencil.

2. Holding the cup over the bucket, cover the hole with your finger and fill the cup halfway with water.

3. Hold the cup over the bucket. Remove your finger and allow the water to pour out of the hole for three seconds. Observe the water as it goes into the bucket.

4. Cover the hole again with your finger. Then remove your finger while you drop the cup into the bucket. Observe the cup and the water.

5. Now drop an empty cup into the bucket at the same time as a full cup. Note the difference in mass. Record your observations on the next page.

🔵 **GO ONLINE** Use the Personal Tutor _Gravity_ to learn about what affects the strength of gravity.

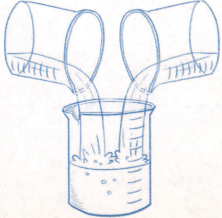

6. **Record Data** Record your observations in the space below. Draw a diagram to help explain what you saw.

7. In what direction does gravity pull objects?

_____

_____

**Talk About It**

Compare your results with a partner's results. Why do you think the water in the cup behaved the way it did when the cup was dropped?

REVISIT Revisit the Page Keeley Science Probe on page 5.
PAGE KEELEY
SCIENCE
PROBES

INQUIRY ACTIVITY

8. Draw a diagram of you and a friend performing this experiment on opposite sides of Earth. Draw arrows to indicate the direction the cups are falling.

 How does Earth's gravity **affect** the Moon? Share your thinking with a partner. **Support your argument** with evidence.

_____

_____

_____

_____

_____

_____

How Could You Become a Physics Teacher?

Physics teachers attend a four-year college and have a degree in physics or education. In addition to studying science, physics teachers may take courses in education. These courses prepare them to teach students and manage a classroom.

Physics teachers are organized and enjoy creating lesson plans that engage all of the students in their classrooms. They have strong communication skills and adapt their teaching to meet the needs of their students. They are able to present science concepts in a way that is easy for students to understand.

It's Your Turn

Some physics teachers teach aspiring astronauts in their classroom. What information would you learn from a physics teacher if you wanted to become an astronaut?

Review

EXPLAIN
THE PHENOMENON

What caused this crater to form?

Summarize It

Use what you have learned to explain what pulls objects toward Earth's surface.

_____

_____

_____

_____

_____

_____

_____

_____

_____

_____

REVISIT
PAGE KEELEY
SCIENCE PROBES

Revisit the Page Keeley Science Probe on page 5. Has your thinking changed? If so, explain how it has changed.

Three-Dimensional Thinking

The Moon orbits around Earth due to gravity. Throughout the Moon's orbit, there appears to be changes in Earth's water level. You have learned that these are called tides.

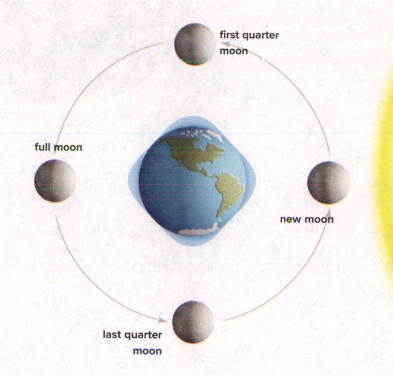

1. Based on what you know about gravity, what causes Earth's changing tides?

   A. Tides are caused by tropical storms such as hurricanes.

   B. Tides are caused by the pull of gravity between Earth and the Moon.

2. What causes the tide to bulge on the side of Earth facing the Moon?

   A. Earth's water bulges on the side facing the Moon because of the pull of gravity.

   B. The Moon reverses its orbit and causes the tides to change.

Extend It

You are a marine biologist studying sand crabs on the beach. In a small group, come up with a plan for the best time to study the crabs based on the tides. Research the times of high tide and low tide. Create a data table for the best time of day to research the crabs. Communicate your ideas to the class.

_____

_____

_____

_____

_____

_____

_____

OPEN INQUIRY

What do you still wonder about the force of gravity and how it affects objects on Earth's surface?

_____

_____

Plan and carry out an investigation to find the answer to your question.

KEEP PLANNING

STEM Module Project
Science Challenge

Now that you have learned about the role of gravity in space, go to your Module Project to explain how the information will affect your model of a planetarium.

Phases of the Moon

Look at the diagram of the Sun, Earth, and Moon. What would a nighttime observer on Earth see? Circle the answer that best matches your thinking.

A.

A full moon

B.

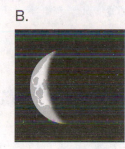

A crescent moon

C.

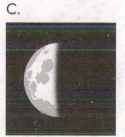

A quarter moon

D.

No moon is seen

Explain why you agree.

_____

_____

_____

_____

You will revisit the Page Keeley Science Probe later in the lesson.

Earth's Motion

How can we use the Sun to tell time?

GO ONLINE

Check out *Sundial* to see the phenomenon in action.

Sundial

Talk About It

Look at the photo and watch the video about the sundial. What questions do you have about the phenomenon? Record or illustrate your thoughts below.

Did You Know?

The first sundials were used in ancient Egypt about 3,000 years ago. They are the oldest known tools for telling time.

Hands On

Shadow Measurements

Think about how a sundial works, and how the Sun changes position in the sky throughout the day. Investigate how shadows are affected by the Sun's changing position.

Make a Prediction How will the length and direction of a shadow change during the day? Explain.

Materials

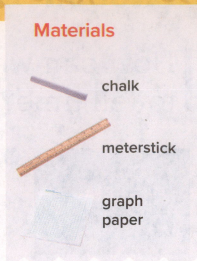

chalk

meterstick

graph paper

_____

_____

_____

Carry Out an Investigation

BE CAREFUL You should never look directly at the Sun.

1. On a sunny morning, go outside to the location your teacher indicates.

2. Note which direction the shadow of a building or tree are stretching. Turn so your shadow is going in the same direction as these objects.

3. With a partner, find some space away from other students. Stand with your back to the Sun.

4. Have your partner trace your feet with the chalk and mark the very end of your shadow. Switch jobs throughout the experiment.

5. **Record Data** Measure the distance between the two chalk marks, and record it in the table below.

6. Repeat these steps at midmorning, midday, and twice in the afternoon. Be sure to place your feet in the outlines from step 4.

Time of Day				
Length of Shadow				

7. **MATH Connection** Use a separate sheet of graph paper to create a bar graph with length of shadow on the vertical axis and time of day on the horizontal axis. Remember to label your graph.

Communicate Information

8. How did the length of your shadow change throughout the day? How did this compare to your prediction?

_____

_____

_____

9. Explain the pattern of change in the length of your shadow.

_____

_____

_____

_____

_____

_____

10. Describe the path of the Sun across the sky during the day.

_____

_____

_____

💬 Talk About It

Compare your results with a partner's results. How could you improve this investigation to collect more evidence to support your prediction?

Earth in Space

VOCABULARY

Look for these words as you read:

moon phases

orbit

revolution

rotation

Earth is moving at 30 kilometers/second (19 miles/second) as it orbits the Sun. A **revolution** is a complete pass around the Sun, taking 365 ¼ days, or one year. Earth is also spinning on its axis at about 1,600 kilometers/hour (1,000 miles per hour). The dotted line through the center of Earth in the image below is its axis. One **rotation** is a complete spin on the axis. Earth makes one rotation every day or every twenty-four hours. Living things do not feel these movements because they are moving with Earth.

Earth's Rotation

At any point in time, half of Earth's surface faces the Sun and is in daylight. The other half of Earth's surface faces away from the Sun and is in darkness.

The tilt of Earth's axis affects the length of the day. If the axis were not tilted, day and night would each be twelve hours long. Instead, there are more hours of daylight and fewer hours of darkness during the summer. In the winter, the amount of daylight is shorter.

This diagram shows Earth's rotation and axis. The green arrow represents the direction of Earth's rotation, while the dotted line shows Earth's tilt, or axis.

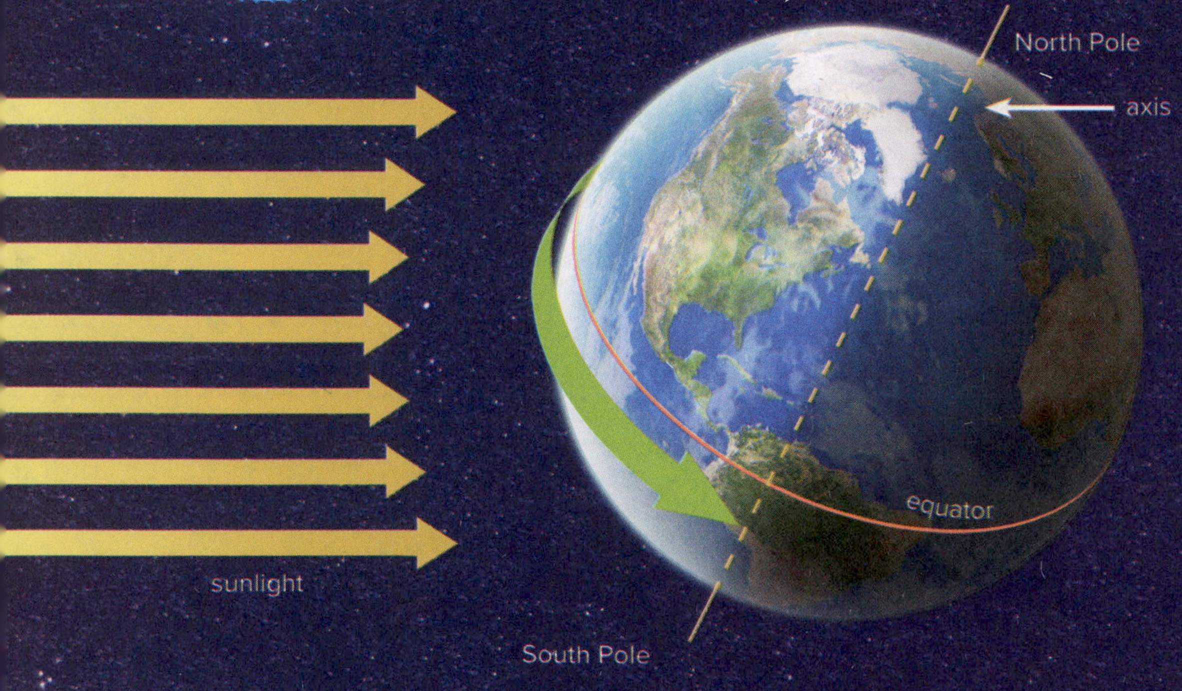

North Pole

axis

equator

sunlight

South Pole

Copyright © McGraw-Hill Education

Earth's Revolution

Earth revolves around the Sun. To revolve means to move around another object. The path a revolving object follows is its **orbit**. Earth's orbit is shaped like an ellipse, or a slightly flattened circle. Earth's orbit around the Sun takes 365 ¼ days, or one year.

Recall that Earth's axis, the imaginary line about which it rotates, is tilted. Earth tilts at an angle of 23.5°. The tilt causes sunlight to strike different parts of Earth at different angles. At any given time, each hemisphere, or half, of Earth gets more or less sunlight than the other. The seasons result from both Earth's tilted axis and its revolution around the Sun.

How Seasons Change in the Northern Hemisphere

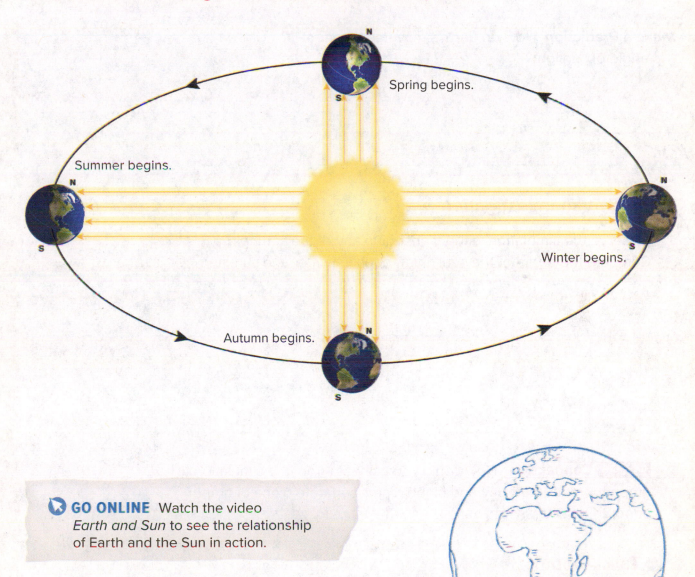

Spring begins.

Summer begins.

Winter begins.

Autumn begins.

GO ONLINE Watch the video *Earth and Sun* to see the relationship of Earth and the Sun in action.

INQUIRY ACTIVITY

Earth's Movements

🔵 GO ONLINE

Investigate the interaction between Earth and the Sun as you conduct the simulation.

Make a Prediction How do Earth's movements affect the angle of sunlight?

_____

_____

_____

Carry Out an Investigation

1. Observe the simulation without changing any of the settings. Compare the angle of the sunlight at noon in winter and summer.

_____

_____

_____

_____

2. **MATH Connection** How could you use data from the simulation to graph the pattern of day and night? Use a separate piece of graph paper to do so.

💬 Talk About It

Did the results support your prediction? Talk about it with a partner.

Seasons

As Earth revolves around the Sun, the tilted axis always points in the same direction. When the Northern Hemisphere tilts away from the Sun, the Northern Hemisphere's surface does not receive as much energy, and temperatures are lower. In the Northern Hemisphere, this is winter.

At the same time, it is summer in the Southern Hemisphere. The Southern Hemisphere tilts toward the Sun, so Sun's energy is more concentrated. The surface receives more energy, and temperatures are warmer.

Because the tilt of Earth's axis always points in the same direction, the seasons in the Northern Hemisphere and the Southern Hemisphere are always opposite. In spring and autumn, both hemispheres receive equal warmth from the Sun, making temperatures similar in both hemispheres.

Winter
about December 21–March 20

Spring
about March 20–June 21

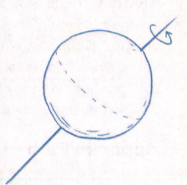

Summer
about June 21–September 22

Autumn
about September 22–December 21

Earth's Revolution—Seasons and the Sun

The Sun's position in the sky appears to change from season to season. Although the Sun does not move, the diagram below shows the Sun's apparent path across the sky during the day as Earth rotates. Each yellow circle represents the Sun's position at midday. The Sun rises much higher in the sky during a summer day. The day on which the Sun appears highest in the sky is known as the summer solstice. In the Northern Hemisphere, the summer solstice occurs around June 21 each year. During this time of year, the Northern Hemisphere tilts more toward the Sun.

In winter, the Sun appears much lower in the sky. In the Northern Hemisphere, the winter solstice occurs around December 21. This is the day on which the Sun appears lowest in the sky. At this time, the Northern Hemisphere tilts away from the Sun.

Halfway between the solstices, neither hemisphere is tilted toward the Sun. The noon Sun is almost directly overhead. Each of these days is known as an equinox. During an equinox, day and night are each about twelve hours long. In the Northern Hemisphere, the spring, or vernal, equinox occurs around March 21. The fall, or autumnal, equinox occurs around September 22.

GO ONLINE Explore *Daylight and Seasons* to analyze data of the amount of daylight throughout the year.

Apparent Path of the Sun

Label the season in which the Sun follows each path.

Copyright © McGraw-Hill Education Fotosearch Premium/Getty Images

Cut out the Notebook Foldables given to you by your teacher. Glue the anchor tabs as shown below. Use what you have learned to show how Earth's tilt on its axis causes the seasons.

Glue anchor tab here.

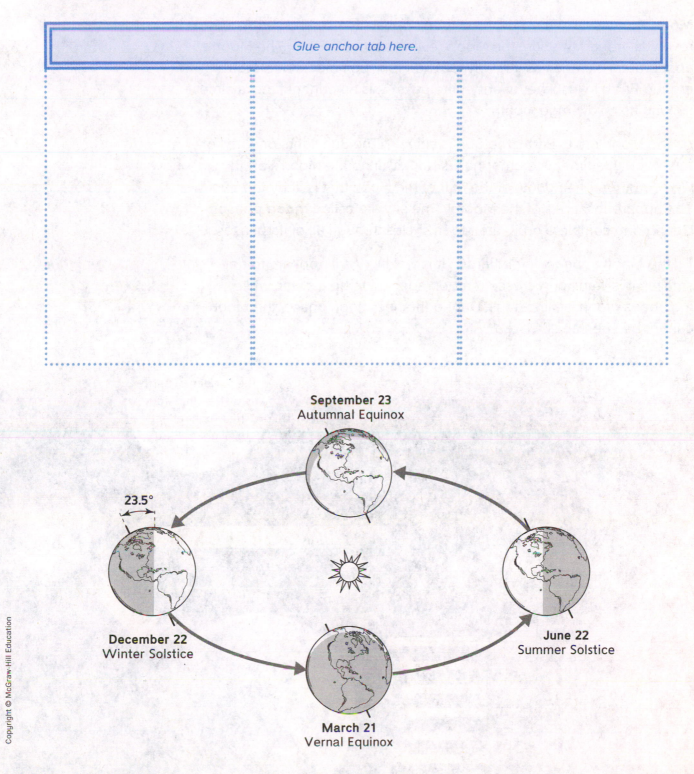

September 23
Autumnal Equinox

23.5°

December 22
Winter Solstice

March 21
Vernal Equinox

June 22
Summer Solstice

Earth's Moon

On many nights, the Moon appears to be the largest, brightest object in the sky. Unlike stars, however, the Moon does not make its own light. Instead, it reflects the light of the Sun.

Moon Phases Like the Sun, the Moon appears to rise and set. As Earth revolves around the Sun, the Moon revolves around Earth. The Moon's appearance changes as it revolves. The Moon completes one orbit around Earth in just over twenty-nine days. This amount of time is almost as long as an average month.

As the Moon orbits Earth, the Sun is shining. The Sun lights one half of the Moon at a time. The other half is dark. During the Moon's orbit, we see different amounts of the half of the Moon that is lit by the Sun. The apparent shapes of the Moon in the sky are called **moon phases**. During one complete orbit, the Moon cycles through all of its phases.

As the Moon appears to get larger, it is waxing. As it appears to get smaller, it is waning. A crescent moon appears to be a sliver, while a gibbous moon is almost full. During the new moon phase, the Moon cannot be seen at all.

Sunlight strikes the surface of Earth as well as the Moon. The Moon reflects this light to Earth.

Sun

sunlight

Earth

Moon

Phases of the Moon

third quarter moon
The Moon is three quarters of the way around Earth.

waning crescent moon
The left sliver of the Moon is the only part you can see.

waning gibbous moon
Slightly less of the lit side can be seen.

new moon
The lit side cannot be seen from Earth.

full moon
The entire lit side can be seen.

waxing crescent moon
Some of the lit side can be seen.

waxing gibbous moon
The Moon is almost full.

first quarter moon
The Moon is a quarter of the way around Earth.

The "mini-moons" along the blue inner circle indicate the direction from which the Sun's rays are shining.

REVISIT PAGE KEELEY SCIENCE PROBES Revisit the Page Keeley Science Probe on page 23.

Data Analysis

Three Cities

Think about Earth's motion and how it causes changes. Different locations on Earth's surface experience these changes in different ways.

Make a Prediction How does Earth's motion affect the average temperatures around the world?

_____

_____

_____

Carry Out an Investigation

1. Research the average high temperature of three different cities around the world. Record the data in the tables.

City 1 Juneau, Alaska	Jan	Feb	Mar	Apr	May	Jun	Jul	Aug	Sep	Oct	Nov	Dec
Average High Temperature (°C)												

City 2 Santiago, Galapagos Islands	Jan	Feb	Mar	Apr	May	Jun	Jul	Aug	Sep	Oct	Nov	Dec
Average High Temperature (°C)												

City 3 Adelaide, Australia	Jan	Feb	Mar	Apr	May	Jun	Jul	Aug	Sep	Oct	Nov	Dec
Average High Temperature (°C)												

2. **Analyze Data** Using the space below, create a line graph that shows how the climates of the three cities are similar and different. Use a different colored pencil to represent each city. Make a key for your graphs and label the axes.

INQUIRY ACTIVITY

Communicate Information

3. How does the timing of the seasons compare in cities north and south of the equator?

_____

_____

_____

_____

4. What relationship did you find among the temperatures in the three cities?

_____

_____

_____

_____

_____

 Analyze the data to explain how the **climate patterns** of each city are based on its location and the **movement of Earth around the Sun**.

_____

_____

_____

_____

💬 Talk About It

Share and discuss your results with your partner and then with the class. Do you notice any patterns among the groups?

A Day in the Life of a Field Engineer

January 5, 1985

As **field engineers**, my coworker and I have to inspect space shuttles for NASA. We enjoy working with aerospace engineers and studying the mechanics of their spacecraft. Today we looked at a space shuttle that was having trouble with its navigation panel. Astronauts rely on this on their space missions. We have had the honor of working with several great astronauts, including Sally Ride, the first American woman to go to outer space.

After inspecting the shuttle, we determined that connection wires to the navigation screen had corroded. We replaced the cables and secured them into place. We are hoping to simulate a test launch with the astronaut to see if the problem is resolved.

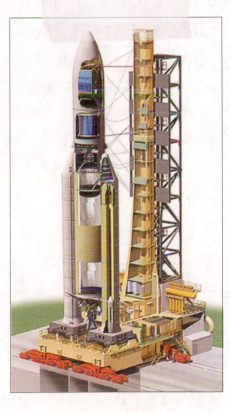

It's Your Turn

Why do you think it is important for field engineers and astronauts to work together?

Review

EXPLAIN
THE PHENOMENON

How does the Sun's position allow us to tell time on a sundial?

Summarize It

Explain how Earth moves through space and how it affects life on Earth.

_____

_____

_____

_____

_____

_____

_____

_____

REVISIT
PAGE KEELEY SCIENCE PROBES

Revisit the Page Keeley Science Probe on page 23. Has your thinking changed? If so, explain how it has changed.

1. Based on what you learned about the patterns of Earth's movement, how does the illustration below show how Earth experiences day and night?

_____

_____

_____

_____

_____

_____

_____

_____

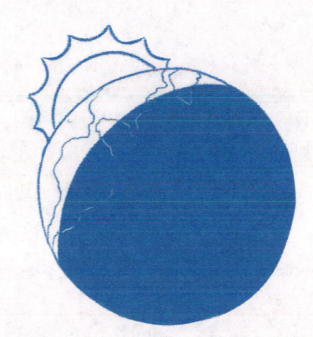

2. The Moon looks completely dark as seen from Earth. What causes a new moon?

   A. The side of the Moon that is lit by the Sun can be seen from Earth.

   B. The side of the Moon that is lit by the Sun cannot be seen from Earth.

Extend It

You are a peer tutor to a student in a younger grade. They ask you what causes day and night to occur. How could you use what you know to design a lesson to teach younger students about Earth's movements?

_____

_____

_____

_____

_____

_____

_____

_____

_____

_____

_____

_____

OPEN INQUIRY

What question do you still have about Earth's motion?

_____

_____

Plan and carry out an investigation to find the answer to your question.

KEEP PLANNING

STEM Module Project
Science Challenge

Now that you have learned about Earth's motion, go to your Module Project to explain how the information will affect your model of a planetarium.

Design a Planetarium Model

PRIMARY SOURCE

Think about how an astronaut collects and communicates information about the movement of Earth. You and your classmates will make a model of a planetarium by choosing a pattern to research to add to the model. Use what you have learned about Earth's motion throughout the module. You will present your information to the class.

Planning after Lesson 1

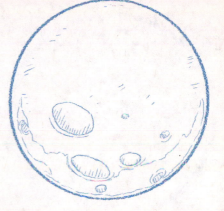

Choose one of the following patterns in our solar system:

1. changes in the length and direction of shadows

2. day and night

3. moon phases

4. seasonal changes

Apply what you have learned about the role of gravity to the pattern of your choosing.

How does gravity affect the Earth pattern that you chose? Explain.

Planning after Lesson 2

Apply what you have learned about Earth's patterns of movement to fill out the graphic organizer below.

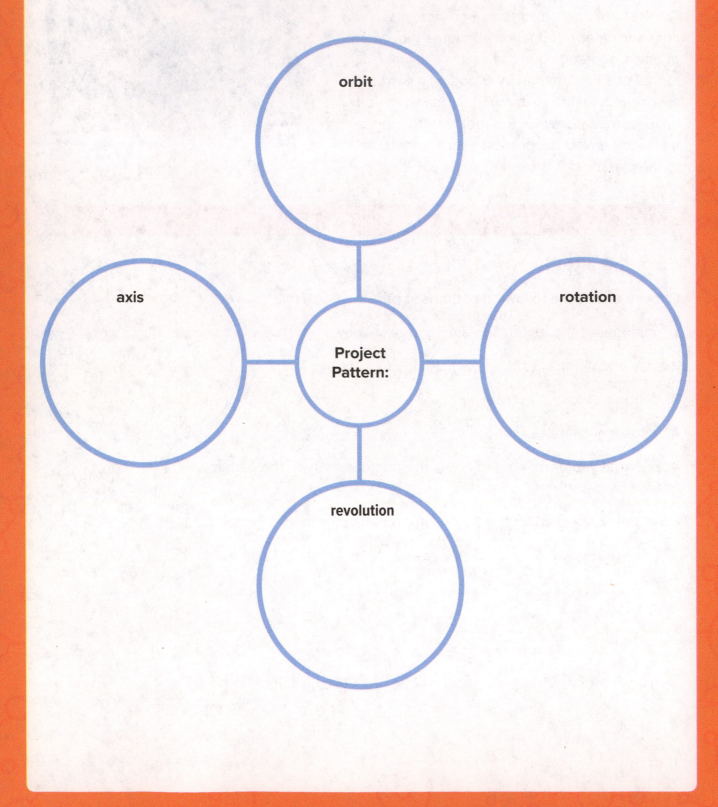

orbit

axis

Project Pattern:

rotation

revolution

Sketch Your Model

Draw your ideas for your model. Select the one that best communicates information and demonstrates the pattern of movement.

Improve Your Model

Share your ideas with a partner. Use what you discuss to improve or refine your model. Record your notes or updated model below.

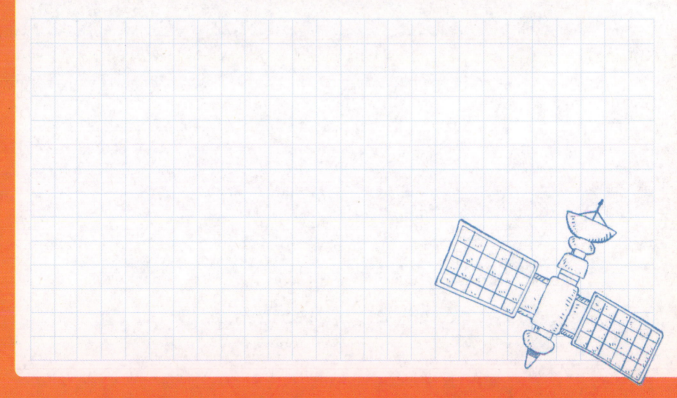

Design a Planetarium Model

Look back at the planning you did after each lesson.
Use that information to complete your final module project.

Materials

Build Your Model

1. Use your project planning to build your model.

2. Determine the materials you will need to build your model. List your materials on the lines.

3. Your model should communicate your research about the pattern you chose.

4. Organize your information to present to the class.

Communicate Your Results

Present your model along with your research and explanation to the class. Write your research presentation on the lines below. Use a separate piece of paper if you need to.

MODULE WRAP-UP

REVISIT
THE PHENOMENON

Using what you learned in this module, explain how the Moon appears to change shape throughout the month.

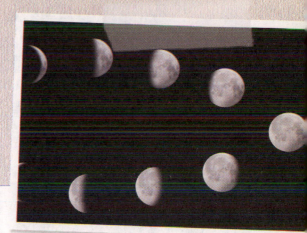

Have your ideas changed? Explain.

_____

_____

_____

Earth and Space

How are stars arranged in the night sky?

The North Star

GO ONLINE

Check out *The North Star* to see the phenomenon in action.

Talk About It

Look at the photo and watch the video of stars around Polaris. What questions do you have about the phenomenon? Talk about them with a partner.

Did You Know?

Polaris, also known as the North Star, cannot be seen at all from Earth's Southern Hemisphere.

Model a Constellation

You will learn about Earth's place in space and star patterns in the night sky. At the end of this module, you will choose one constellation and build a model of how its stars are arranged in the night sky. You will include information about each of the stars that belong to that particular constellation and when during the year you are most likely to see this pattern of stars in the night sky.

Lesson 1
Earth's Place in Space

Lesson 2
Stars and Their Patterns

Aerospace engineers develop airplanes and spacecraft. Christine Bland is an aerospace engineer who has helped build the Orion spacecraft, which was developed to have astronauts travel further into space. When she was young, she enjoyed math and science. These skills helped Christine pursue her goals of working on spacecraft and developing new technology for space travel. The information that is collected by the space shuttles built by aerospace engineers can be used by many other types of space scientists.

STEM Module Project

Plan and Complete the Science Challenge Use what you learn to model a constellation from the perspective of where you are on Earth!

Earth and the Sun

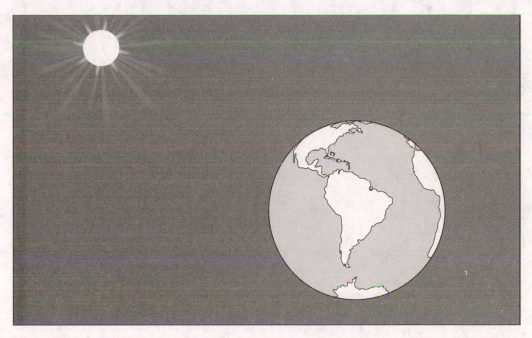

Emmy and Dexter were talking about Earth and the Sun. They each had different ideas about movement in the Earth-Sun system. This is what they said:

Emmy: *I think Earth moves around the Sun.*

Dexter: *I think the Sun moves around Earth.*

Who do you agree with most? _____

Explain why you agree.

_____

_____

_____

_____

_____

You will revisit the Page Keeley Science Probe later in the lesson.

Earth's Place in Space

ENCOUNTER
THE PHENOMENON

What does the sunrise look like where you live?

Sunrise

▶

🔵 GO ONLINE

Check out *Sunrise* to see the phenomenon in action.

💬 Talk About It

Look at the photo and watch the video of the sunrise from space. What questions do you have about the phenomenon? Record or illustrate your thoughts below.

Did You Know?

Astronauts see a sunrise every 92 minutes aboard the International Space Station!

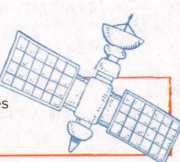

INQUIRY ACTIVITY

Hands On

Model of the Sun, Earth, and Stars

Think about how the sunrise looks from space as compared to where you live and what you know about Earth in space.

Make a Prediction How does Earth's location determine which stars you see throughout the year?

_____

_____

_____

Carry Out an Investigation

1. Turn the plastic cup upside down. Form a ball with the yellow clay. Stick the ball onto the bottom of the plastic cup to represent the Sun.

2. Form a ball with the green clay to represent Earth. Use the brass fastener to carve Earth's hemispheres into the clay.

3. Fold the black construction paper in half, long end to long end, and crease. Unfold the paper and cut the creased line. Tape the two strips together to form one long strip.

4. Use the white crayon to draw stars on one side of the long strip. Vary the size of the stars and the distance they are apart from each other to represent the night sky. Tape the ends of the strip together to form a circle, with the stars facing inward.

5. Insert the wood dowel into the top of your clay model of Earth. Position Earth so it appears tilted on its axis.

6. Stick the brass fastener into a location on Earth.

7. Place the Sun in the center of the night sky. Using what you know about Earth's movements, hold the wood dowel to move Earth in its rotation and revolution.

Materials

plastic cup

yellow modeling clay

green modeling clay

brass fastener

black construction paper

scissors

white crayon

masking tape

wood dowel rod

Communicate Information

8. Draw how your model shows the relative position and motion of Earth and the Sun.

9. How does your model show how Earth and the stars interact in space?

_____

_____

_____

10. The Sun and stars in this model remained in the same location while Earth moved. What can you infer about the Sun?

 Use this model to **support an argument** that object's in space **vary in size.**

_____

_____

_____

💬 **Talk About It**

What do you wonder about the parts of space visible from where you live? Talk about it with a partner.

VOCABULARY

Look for these words as you read:

apparent motion

galaxy

planet

Apparent Motion of Stars

The stars in the northern sky seem to circle around Polaris, which is also known as the North Star. **Apparent motion** is the way something appears or seems to move. The stars appear to move because of Earth's rotation and revolution. Although the stars appear to change position, their position in the universe sky does not change.

Different stars are visible at different times of the year depending on the location of an observer on Earth's surface. This is due to Earth's revolution around the Sun. The stars that are visible in a certain location on Earth's surface changes slowly throughout the year. Each night, the position of most stars shifts slightly to the west. As the stars once visible in the west leave our view, other stars appear in the east.

The night sky as seen from this location on Earth will look different at another time during the year.

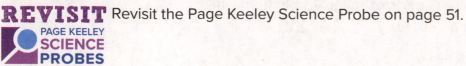

REVISIT Revisit the Page Keeley Science Probe on page 51.

PAGE KEELEY SCIENCE PROBES

Our Galaxy

Our location in space is part of a larger group of stars called a galaxy. A **galaxy** contains billions of stars, dust, and gas that are held together by gravity. Astronomers estimate there are 100 billion galaxies in the universe. Our galaxy is known as the Milky Way. It gets its name from the ancient Greeks, who called the streaks of light they saw in the sky "milky circle." Ancient Romans called the streaks "road of milk."

▶ **GO ONLINE** Watch the video *Our Place in the Milky Way*. Talk to a partner about the location of the Sun and Earth within the galaxy.

Copyright © McGraw-Hill Education   Las Cumbres Observatory/Doug Shobbrook

Stars in the Milky Way are grouped into four arms that spiral out from the center.

The Milky Way contains more than 200 billion stars. The dust and gas in the galaxy are enough material to make billions of more stars. Astronomers believe the galaxy creates up to seven new stars each year. Even so, the Milky Way is not the largest galaxy.

It is not possible to photograph the entire Milky Way. Astronomers calculate the size and structure of our galaxy using what they know about other galaxies. Just like Earth travels around the Sun, the Sun travels around the center of the Milky Way. It takes 250 million years for the Sun to make one complete trip around the center of the galaxy.

PRIMARY SOURCE

Inspect

Read the passage *Making a Map of the Night Sky*. Underline text evidence that tells you how William Herschel determined the location and shape of the Milky Way galaxy.

Find Evidence

Reread the third and fourth paragraphs. Where is the Sun located in the Milky Way?

Highlight text evidence that supports your answer.

Notes

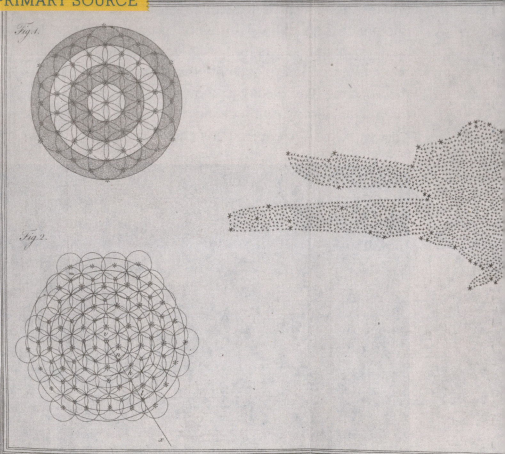

Making a Map of the Night Sky

By the 1700s, astronomers had become experts at charting the position of the night sky's brightest stars. Their two-dimensional models of the sky were very practical. Astronomers continued to navigate the stars visually. By mapping more and more stars, they could further understand the universe.

Some astronomers wanted to start looking at the larger picture of the sky. They began thinking of mapping the three-dimensional universe. One focus was to explain the location and shape of the Milky Way. Sir William Herschel built on previous knowledge and counted the number of stars in almost 700 different regions of the sky. He estimated the distance between the stars and Earth.

From these counts and estimates, Herschel could begin working on a model of the galaxy. The result of Herschel's work was a three-dimensional model showing the Milky Way, in which he believed the Sun to be the center.

We now know that the Sun does not appear in the center of the Milky Way. Later astronomers were able to use Herschel's model and further research to understand that the Sun is located within one of the Milky Way's spiral arms.

How did Hershel use math to make the map of the Milky Way?

_____

_____

_____

_____

Make Connections
💬 Talk About It

What do you wonder about how scientists count the stars today as compared to in the past? Talk about it with a partner. Make a plan to research your questions.

Notes

_____

_____

_____

_____

_____

_____

_____

_____

_____

_____

_____

The Solar System

Within the Milky Way galaxy is our solar system, which consists of the Sun and all of the objects that orbit around it. One type of object that orbits the Sun are planets. A **planet** is a large, round object in space that orbits a star.

Planets of the Solar System From nearest to farthest from the Sun, the planets in our solar system are Mercury, Venus, Earth, and Mars, or the inner planets. Next are Jupiter, Saturn, Uranus, and Neptune, or the outer planets. The planets revolve in elliptical, or nearly circular, orbits around the Sun. Several planets are visible in the night sky from Earth from time to time, even without a telescope. Visible planets include Mercury, Venus, Mars, Jupiter, and Saturn. Planets do not make their own light, but reflect the light from the Sun.

Between the inner and outer planets is a belt of space rocks called asteroids. These are rocky or metallic objects that also orbit the Sun within the solar system.

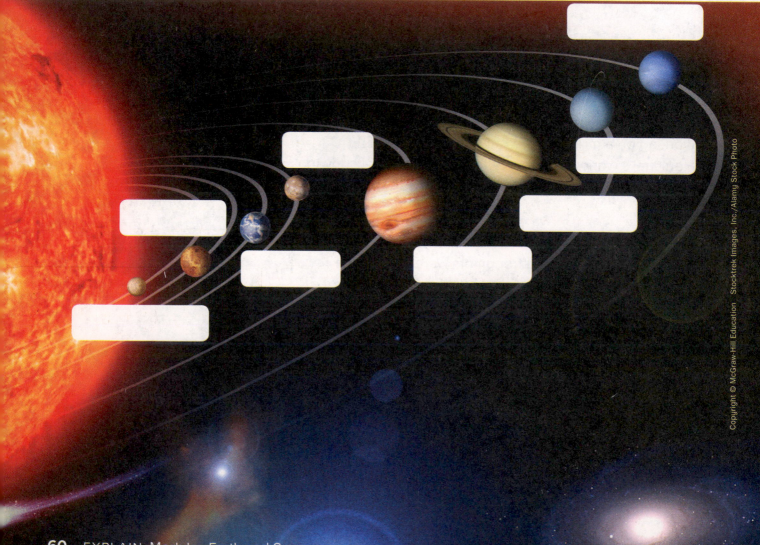

INQUIRY ACTIVITY

Hands On

Model the Solar System

You are going to use a model of the solar system to help you explain more about Earth's location in outer space.

State the Claim Where is Earth located in the solar system?

_____

_____

Materials

scissors

construction paper

3.5 meters of string

meterstick

Make a Model

1. Cut circles from construction paper to represent the Sun, the inner planets, and the outer planets.

2. Lay the string out in a straight line on the floor. Place the Sun at one end of the string.

3. The distances of each planet from the Sun have been converted to centimeters in the table below. Place each planet at the correct distance from the Sun on the string.

Space Object	Distance from the Sun (cm)
Mercury	4 cm
Venus	7 cm
Earth	10 cm
Mars	15 cm
Jupiter	52 cm
Saturn	96 cm
Uranus	192 cm
Neptune	300 cm

💬 Talk About It

Does the activity support your claim? Talk about it with a partner.

A Day in the Life of an Astronomer

Interview with Edward Gomez, astronomer

How would you describe your career in one to three sentences?

I use math, computer coding, and science to look for asteroids that pass close to Earth. I also create interactive websites for educational projects aimed at students, using robotic telescopes.

How do you use science and engineering in your career?

I use science to hunt and track asteroids that are speeding past Earth so that we can follow their positions and movements. We take a lot of images of the asteroids to learn more about what they are made of, how fast they are moving, and how fast they spin.

Describe what a typical day looks like. What does an exciting day look like?

A typical day involves emailing people located around the world. I coordinate

projects with people in the United States, Australia, and Europe. Normally, I spend time looking at astronomical images and writing Python code to analyze the images. An exciting day involves doing a lot of data analysis, coding, and writing a press release about a new discovery in science.

It's Your Turn

What do you wonder about the role of computer coding in studying space? Research how knowing how to code can help you reach your career goals.

WRITING Connection Use what you have learned to describe your location within the universe, starting with your class room number. Include as many details as you can. Write your response on the lines below and use your response to present your location. Listen carefully as others present their location to you as well.

_____

_____

_____

_____

_____

_____

_____

_____

_____

_____

Draw a diagram or illustration that supports your written description of your location within the universe.

Review

EXPLAIN
THE PHENOMENON

What does the sunrise look like where you live?

Summarize It

Explain Earth's location in space and how its location affects what we see in the night sky.

_____

_____

_____

_____

_____

_____

_____

_____

_____

REVISIT
PAGE KEELEY SCIENCE PROBES

Revisit the Page Keeley Science Probe on page 51. Has your thinking changed? If so, explain how it has changed.

Three-Dimensional Thinking

1. Based on the data table, what conclusion can you draw?

Planet	Length of Day (hours)	Length of Year (Earth years)	Distance from the Sun (AU)
Mercury	1,408	0.2	0.4
Venus	5,832	0.6	0.7
Earth	24	1.0	1.0
Mars	25	1.9	1.5
Jupiter	10	11.9	5.2
Saturn	10	29.4	9.5
Uranus	17	84.0	19.2
Neptune	16	164.8	30.0

A. The farther a planet is from the Sun, the longer its day.

B. The farther a planet is from the Sun, the longer its year.

C. A day on Earth is longer than a day on Venus.

D. Uranus is the coldest of all the planets.

2. A planet is a large, round space object that _____ the Sun.

A. attracts   C. follows

B. orbits   D. reflects

3. Circle all that apply.

Stars appear to move in the sky because of Earth's _____.

A. axis

B. rotation

C. poles

D. galaxies

E. revolution

Extend It

Think about how different types of technology allow space scientists to explore space. How have new types of technology helped space scientists, such as astronomers, have a better understanding of the universe? Research different types of technology to include examples.

_____

_____

_____

_____

_____

_____

_____

_____

_____

_____

_____

OPEN INQUIRY

What do you still wonder about the organization of objects in space?

_____

_____

_____

Plan and carry out an investigation to find the answer to your question.

KEEP PLANNING
STEM Module Project
Science Challenge

Now that you have learned about where Earth is located in space, go to your Module Project to explain how the information will affect your model of a constellation.

LESSON 2 LAUNCH

Constellations

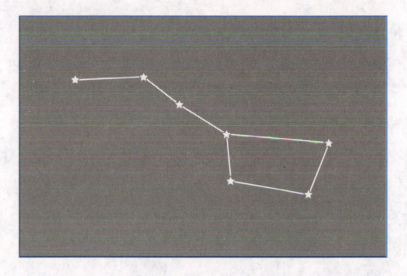

Kendra and Jake were looking at the night sky. They recognized the Big Dipper, which is part of a constellation. They each had different ideas about the constellations. This is what they said:

Jake: *The stars form the same constellations throughout the year. They keep the same patterns even though we see different constellations at different times of the year.*

Kendra: *The stars form different constellations throughout the year. They are always changing their patterns, which is why we see different constellations at different times of the year.*

Who do you agree with more? _____

Explain why you agree.

_____

_____

_____

_____

You will revisit the Page Keeley Science Probe later in the lesson.

Stars and Their Patterns

ENCOUNTER
THE PHENOMENON

Why are some stars brighter than others?

GO ONLINE

Check out *Night Sky* to see the phenomenon in action.

Look at the photo and watch the video about the stars in the sky. What questions do you have about the phenomenon? Talk about them with a partner. Record or illustrate your thoughts below.

Did You Know?

Most stars have a twin! Many stars exist in pairs, meaning they share a center of gravity.

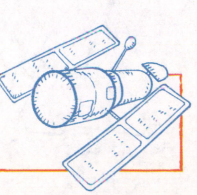

INQUIRY ACTIVITY

Star Brightness

Materials

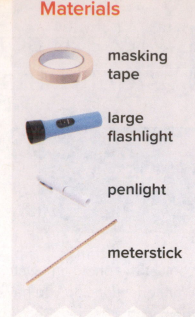

masking
tape

large
flashlight

penlight

meterstick

Think about the night sky. All stars are not the same brightness, although some appear to be similar. In this activity, we will be looking at one of several factors that affect the apparent brightness of a star.

Make a Prediction How does a star's distance from Earth affect how bright it appears?

_____

_____

_____

Carry Out an Investigation

1. Place a piece of tape on the floor marking a distance of 0.5 meters from the wall. Then, leave a tape mark at 2 meters and 4 meters.

2. Stand at the 0.5 meter mark. Shine your flashlight at the wall. Observe how bright the light looks.

3. Repeat steps one and two at a distance of 2 meters and 4 meters from the wall.

4. **Record Data** Describe the brightness of the light as you moved closer and further from the wall. Record your observations in the data table.

5. Now repeat steps 1–3 using the penlight. Record your observations in the data table.

	Flashlight	Penlight
0.5 m from wall		
2 m from wall		
4 m from wall		

Communicate Information

6. Did the amount of light put out by the model star change? Explain.

_____

_____

_____

7. What could you do to get the two "stars" to have the same brightness using the large and small flashlight?

_____

_____

_____

8. What other factors might affect the brightness of a flashlight? What other factors do you think might affect the brightness of a star?

_____

_____

_____

INQUIRY ACTIVITY

9. Imagine that you see two equally bright lights on the wall, but cannot see the flashlights. Do you have enough information to know if the flashlights were equally bright at the same distance or two unequally bright flashlights at different distances?

_____

_____

_____

_____

10. You see two unequally bright spots on the wall. What inferences could you make about the light sources of those spots?

_____

_____

_____

_____

 What **patterns** did you notice in the data collected during this investigation? Share your observations with a partner. **Support your argument** with evidence.

_____

_____

_____

_____

_____

MAKE YOUR CLAIM

Which star is the brightest star seen from Earth?

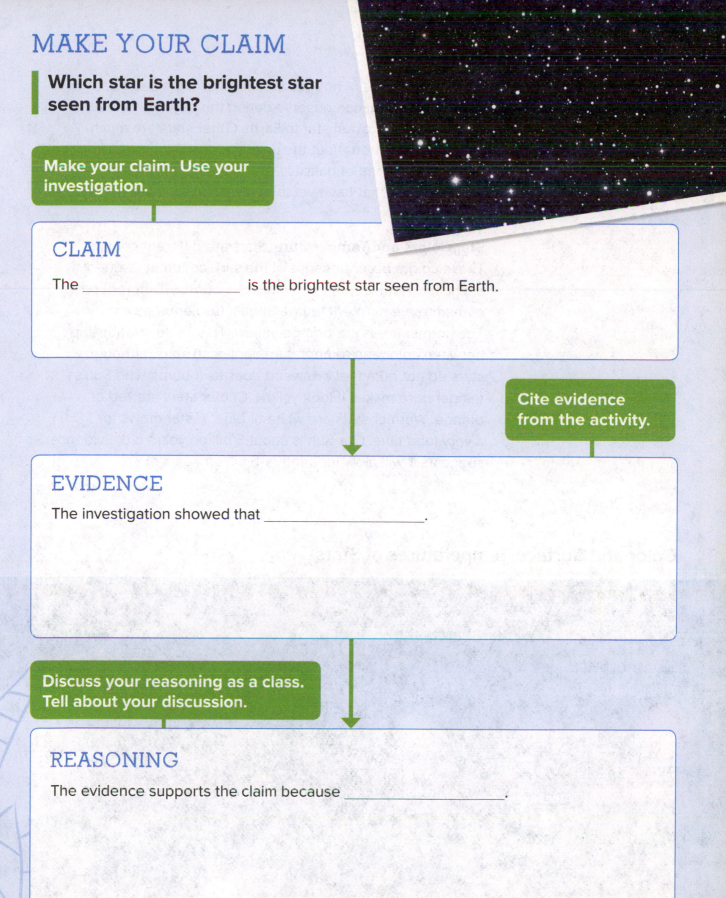

Make your claim. Use your investigation.

CLAIM

The _____ is the brightest star seen from Earth.

Cite evidence from the activity.

EVIDENCE

The investigation showed that _____.

Discuss your reasoning as a class. Tell about your discussion.

REASONING

The evidence supports the claim because _____.

You will revisit this claim throughout the lesson.

Stars

A **star** is a sphere of hot gas that gives off light and heat. The only star you can observe during the daytime is the Sun. The Sun is the closest star to Earth. Other stars are much farther away. Throughout the universe, stars are found in large groups in the form of galaxies. The universe may have many more galaxies that have yet to be discovered, each with billions of stars.

Star Colors and Temperature Stars are different colors. These colors occur because of the surface temperature of each star. Think about the flames of a bonfire. Different parts of the fire are different temperatures. Cooler areas are red. The hottest areas are orange-yellow. This same relationship between color and temperature applies to stars, although stars do not burn fuel like wood does as it burns. The Sun's temperature makes it look yellow. Cooler stars are red or orange. Warmer stars are white or blue. A star glows for a very long time. Our Sun is about 5 billion years old. Evidence suggests it will glow for another 5 billion years or so.

Color and Surface Temperatures of Stars

Betelgeuse
red
3,000°C

Aldebaran
red-orange
5,000°C

Sun
yellow
6,000°C

Altair

Spica
blue
35,000°C

Star Distances The Sun is about 150 million kilometers (93 million miles) from Earth. It takes about eight minutes for its light to reach Earth. Most stars are much farther away. Writing their distances in kilometers becomes difficult to understand. To simplify the writing of such large distances, astronomers use a unit called a light-year. A ==light-year== is the distance light travels in one year, which is nearly 10 trillion kilometers (6 trillion miles). When you observe a distant star, you are actually seeing what it looked like in the past. A star you see today may have stopped glowing many years ago. However, it is so far away that its light is still traveling through space.

Nearest Stars to Earth

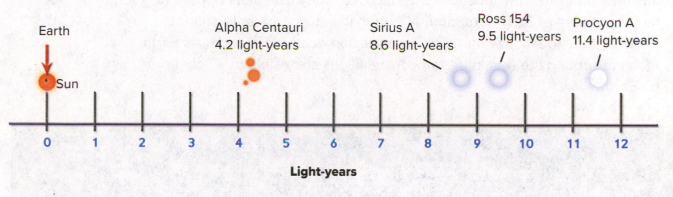

Light-years

Star Cycles Stars form when matter comes together and starts to give off energy. Stars go through stages, or cycles, between their beginning and ending. Different kinds of stars have different cycles. A star's cycle ends when it stops giving off energy. Some cycles, like the one the Sun is going through, take about 10 billion years. Because the Sun is about 5 billion years old, it is about halfway through the cycle.

> **GO ONLINE** Watch the video *Star Cycles* to learn more about the life cycle of stars.

COLLECT EVIDENCE

Revisit your claim on page 73. Add evidence to your claim about the apparent brightness of stars.

Constellations

When people in ancient cultures studied the night sky, they saw patterns in the stars. These patterns are called **constellations**. They were named after animals, fictional characters, and objects.

Star patterns have been useful to both ancient and modern travelers. If you can see either the Ursa Major and Ursa Minor constellations in the night sky, which contain the Big Dipper and Little Dipper, you can use them to easily find Polaris, the North Star. If you travel in the direction of Polaris, you will be moving north.

Today, astronomers divide the sky into 88 constellations. Many of the ancient names for constellations are still used today. The stars appear to move because of Earth's rotation. Although the stars appear to change position, their location within their constellation does not change. As Earth revolves around the Sun, different constellations are visible to an observer on Earth.

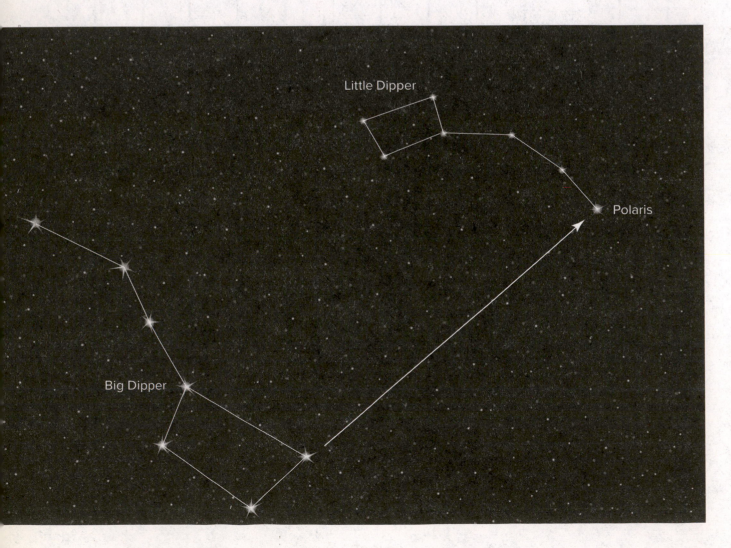

There are a few constellations that are visible all year long in the Northern Hemisphere. These are the ones that are closest to Polaris. As Earth rotates, these constellations seem to circle the North Star. They never appear to rise or set. These constellations are known as circumpolar constellations.

Constellations make sense only to the observers on Earth. Stars that seem close together are actually far apart. If you looked at the location of constellations from a different part of the universe, those patterns would change.

GO ONLINE Explore *Constellations* to learn about star patterns in each hemisphere.

PRIMARY SOURCE

Modern star map

REVISIT Revisit the Page Keeley Science Probe on page 67.

PAGE KEELEY SCIENCE PROBES

INQUIRY ACTIVITY

The Night Sky

🚀 GO ONLINE

Investigate how Earth and the stars change with the seasons by conducting the simulation *The Night Sky*.

State the Claim What happens to the position of stars as the months change?

_____

_____

_____

_____

Carry Out an Investigation

1. What happens to the position of the stars from evening to the next morning?

_____

_____

_____

_____

2. What happens to the position of individual stars over the months of the year? Explain in terms of Earth's position relative to those stars.

_____

_____

_____

_____

_____

💬 Talk About It

Did what you observe in the simulation support your claim?
Talk about it with a partner.

Look at the historical star map below. Compare it to the modern star map on page 77. What is similar and different about the two maps? Talk about it with your class.

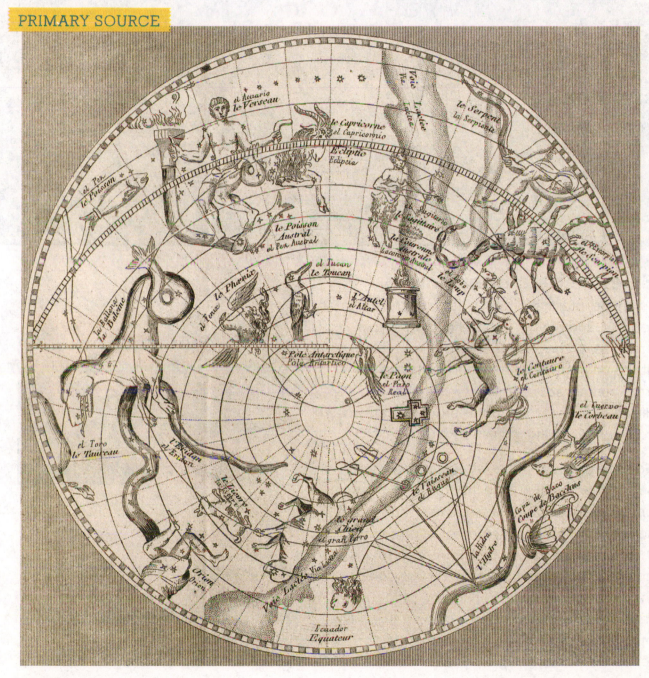

Historical star map

A Day in the Life of a Science Data Operator

Science data operators take

scientific information, such as data and reports, and enter it into the computer to organize it. They spend a lot of time typing at the computer to compile information from various sources. They work with different types of computer programs and are comfortable using technology.

Science data operators work in offices and spend most of their time inputting information on a computer. They work closely with scientists to collect and gather information for their assigned tasks. The information that is entered into the computer is shared within the company or published for the public to read. By organizing this information, they help other scientists learn and advance their research. Science data operators work for all types of companies. Some even work for NASA! If you like numbers and computers and being organized, this career is one you might be interested in.

It's Your Turn

What skills does a science data operator use that could be useful in other careers as well? Talk about it with a partner.

WRITING Connection Choose a constellation to research. Find information about the name of the constellation and the story behind it. Identify where the constellation is visible from Earth. Use the graphic organizer below and a separate piece of paper to write your summary. Draw a sketch of the constellation at the bottom of the page.

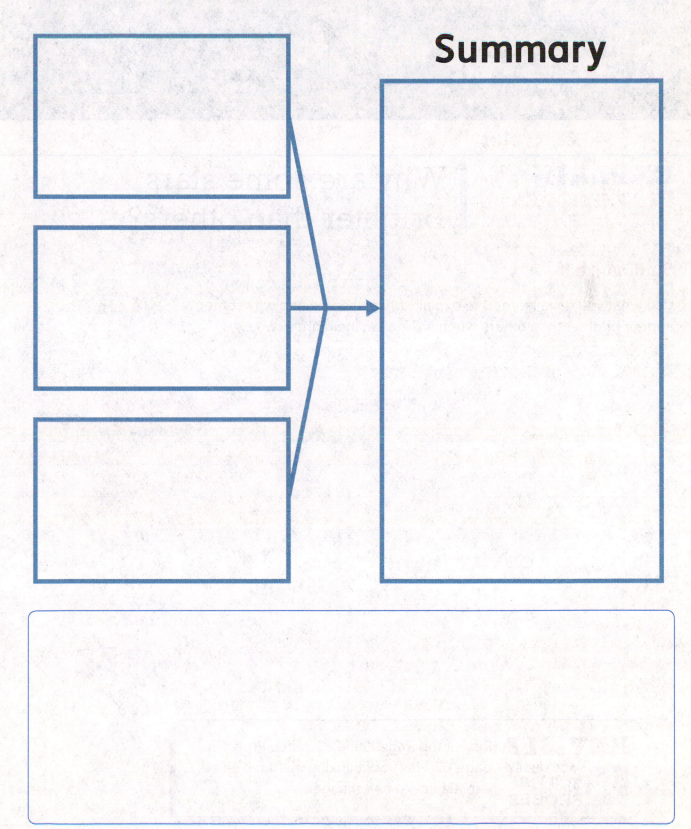

Summary

Review

EXPLAIN
THE PHENOMENON

Why are some stars brighter than others?

Summarize It

Use what you have learned to explain what stars are and why some appear brighter than others when we look at the night sky.

_____

_____

_____

_____

_____

_____

REVISIT
PAGE KEELEY
SCIENCE
PROBES

Revisit the Page Keeley Science Probe on page 67. Has your thinking changed? If so, explain how it has changed.

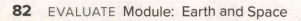

Three-Dimensional Thinking

1. Which statement explains why scientists use a unit called
   a light-year when writing about the large distances between Earth
   and distant stars?

   A. Light travels at different times during different times of the year.

   B. Stars are so far from Earth that writing their distance in kilometers
      can become difficult to understand.

   C. There are too many stars in space to measure using kilometers.

   D. It sounds more scientific.

2. The table shows the distance of five different stars from Earth.

Star	Distance from Earth (light-years)
Star A	8.6
Star B	11.4
Star C	6.0
Star D	4.2
Star E	7.7

   Based on what you learned about star distances, choose the correct
   order of the stars as they appear from brightest to dimmest, based on
   their distance from Earth.

   A. Star B, Star A, Star E, Star C, Star D

   B. Star A, Star B, Star C, Star D, Star E

   C. Star D, Star C, Star E, Star A, Star B

   D. Star E, Star D, Star C, Star B, Star A

Extend It

You have a pen pal who lives in the opposite hemisphere from you. Research the types of constellations that can be seen in your area during the month you are in now. Describe the constellations. Use multimedia to present your research to your pen pal.

_____

_____

_____

_____

_____

_____

_____

_____

_____

_____

_____

_____

_____

_____

_____

KEEP PLANNING
STEM Module Project
Science Challenge

Now that you have learned about stars and star brightness, go to your Module Project to explain how the information will affect your model of a constellation.

Model a Constellation

You have learned about Earth's place in space and star patterns in the night sky. Using what you have learned throughout the module, you will choose one constellation and build a model of how its stars are arranged in the night sky. You will include information about each of the stars that belong to that particular constellation and when during the year you are most likely to see this pattern of stars in the night sky.

Planning after Lesson 1

Apply what you have learned about Earth's location in space to your project planning.

Choose a constellation to research. How does Earth's location determine what the constellation looks like from Earth?

Planning after Lesson 2

Apply what you have learned about stars and their patterns to your project planning.

Record research of the constellation you chose on a separate piece of paper. How could you build a three-dimensional model of a constellation?

Sketch Your Model

Draw your ideas for your constellation model. Select the best idea to build and use to communicate your information.

Model a Constellation

Look back at the planning you did after each lesson.
Use that information to complete your final module project.

Build Your Model

1. Use your project planning to build your model.

2. Determine the materials you will need to build your model. List your materials on the lines.

3. Your model should communicate your research about the constellation you chose.

4. Organize your information to present to the class.

Draw your final model of your constellation. Label the parts of your model to show how it is three-dimensional.

Materials

_____

_____

_____

_____

STEM Module Project
Science Challenge

Communicate Your Results

Present your model along with your research and explanation to the class. Write your research presentation on the lines below.

MODULE WRAP-UP

REVISIT
THE PHENOMENON

Using what you learned in this module, explain how the brightness and patterns of stars are due to their distance from Earth and position in the night sky.

Have your ideas changed? Explain.

Science Glossary

A

abiotic factor a nonliving part of an ecosystem

acid rain harmful rain caused by the burning of fossil fuels

air mass a large region of air that has a similar temperature and humidity

algae bloom a sometimes harmful rapid increase in the amount of algae found in water

apparent motion when a star or other object in the sky seems to move even though it is Earth that is moving

atmosphere the gases that surround Earth

B

bacteria a type of single cell organism

biosphere the part of Earth in which living things exist and interact

biotic factor a living thing in an ecosystem, such as a plant, an animal, or a bacterium

C

chemical change a change that produces new matter with different properties from the original matter

chemical property a characteristic that can only be observed when the type of matter changes

climate the average weather pattern of a region over time

colloid a type of mixture in which the particles of one material are scattered through another without settling out

condensation the process through which a gas changes into a liquid

conductivity ability for energy, such as electricity and heat, to move through a material

conservation the act of saving, protecting, or using resources wisely

conservation of mass a physical law that states that matter is neither created nor destroyed during a physical or chemical change

constellation any of the patterns of stars that can be seen in the night sky from here on Earth

consumer an organism that cannot make its own food

D

decomposer an organism that breaks down dead plant and animal material

deforestation the removal of trees from a large area

deposition the dropping off of eroded soil and bits of rock

E

endangered when a species is in danger of becoming extinct

energy the ability to do work or change something

energy flow the movement of energy from one organism to another in a food chain or food web

erosion the process of weathered rock moving from one place to another

evaporation a process through which a liquid changes into a gas

extinct when a species has died out completely

F

floodplain land near a body of water that is likely to flood

food chain the path that energy and nutrients follow among living things in an ecosystem

food web the overlapping food chains in an ecosystem

fungi plant-like organisms that get energy from other organisms which may be living or dead

G

galaxy a collection of billions of stars, dust and gas that is held together by gravity

gas a state of matter that does not have its own shape or definite volume

geosphere the layers of solid and molten rock, dirt, and soil on Earth

glacier a large sheet of ice that moves slowly across the land

gravity the force of attraction between any two objects due to their mass

groundwater water stored in the cracks and spaces between particles of soil and underground rocks

H

habitat a place where plants and animals live

hot spot an area where molten rock from within the mantle rises close to Earth's surface

hydrosphere Earth's water, whether found on land or in oceans, including the freshwater found underground and in glaciers, lakes, and rivers

I

ice caps a covering of ice over a large area such as in the polar regions

invasive species an organism that is introduced to a new ecosystem and causes harm

L

landslide the sudden movement of rocks and soil down a slope

light year the distance light travels in a year

liquids a state of matter that has a definite volume but no definite shape

M

magnetism the ability of a material to be attracted to a magnet without needing to be a magnet themselves

mass the amount of material in an object

matter anything that has mass and takes up space

meteor a chunk of rock from space that travels through Earth's atmosphere

meteorite A meteor that strikes Earth's surface

minerals solid, nonliving substances found in nature

mixture a physical combination of two or more substances that are blended together without forming new substances

molten rock very hot melted rock found in Earth's mantle

moon phases the apparent shapes of the Moon in the sky

N

nitrogen cycle the continuous circulation of nitrogen from air to soil to organisms and back to air or soil

O

orbit the path an object takes as it travels around another object

oxygen-carbon cycle the continuous exchange of carbon dioxide and oxygen among living things

P

phloem the tissue through which food from the leaves moves throughout the rest of a plant

physical change a change of matter in size, shape, or state that does not change the type of matter

physical property a characteristic of matter that can be observed and or measured

planet a large, round object in space that orbits a star

precipitation water that falls from clouds to the ground in the form of rain, sleet, hail, or snow

predator an animal that hunts other animals for food

prey animals that are eaten by other animals

producer an organism that uses energy from the Sun to make its own food

R

reflectivity the way light bounces off an object

reservoir an artificial lake built for storage of water

revolution one complete trip around an object in a circular or nearly circular path

rotation a complete spin on an axis

runoff excess water that flows over Earth's surface from a storm or flood

S

solid a state of matter that has a definite shape and volume

solubility the maximum amount of a substance that can be dissolved by another substance

solution a mixture of substances that are blended so completely that the mixture looks the same everywhere

star an object in space that produces its own energy, including heat and light

stomata pores in the bottom of leaves that open and close to let in air or give off water vapor

storage the process of water being stored on Earth's surface in the ground or as a water feature

T

tides the regular rise and fall of the water level along a shoreline

transpiration the release of water vapor through the stomata of a plant

V

volcano an opening in Earth's surface where melted rock or gases are forced out

volume a measure of how much space an object takes up

W

water cycle the continuous movement of water between Earth's surface and the air, changing from liquid into gas into liquid

weather the condition of the atmosphere at a given place and time

X

xylem the plant tissue through which water and minerals move up from the roots

Index

A

Aerospace engineers, 50
Apparent motion
  defined, 56
  of stars, 56
Asteroids, 60
Astronauts, 4
Astronomers, 58, 62

B

Big Dipper constellation,
  67, 76

C

Circumpolar constellations,
  77
Constellations, 67
  Big Dipper, 67, 76
  circumpolar, 77
  defined, 76
  Little Dipper, 76

E

Earth
  orbit, 29
  revolution of, 28, 29
  rotation of, 28
  in space, 28
Earth's gravitational pull,
  13
  meteorites and, 14
  meteors and, 14
  Moon and, 13
  tides and, 13

Engineers
aerospace, 50
field, 39

F

Field engineers, 39

G

Galaxy
  defined, 57
  Milky Way, 57
Gomez, Edward, 62
Gravitational pull, 13–14
Gravity, defined, 12

H

Hershel, Sir William, 58–59

L

Light-year, 75
Little Dipper constellation,
  76

M

Milky circle, 57
Milky Way, 57
Moon phases, 34–35
Moon's gravitational pull,
  13
  Earth and, 13
Motion, apparent. *See*
  Apparent motion

N

Northern Hemisphere, 31
North Star, 56, 76, 77

O

Orbit, of Earth, 29

P

Physics teachers, 19
Planet(s)
  defined, 60
  of solar system, 60
Polaris, 76, 77

R

Revolution of Earth, 28, 29
Road of milk, 57
Rotation of Earth, 28

S

Science data operators, 80
Seasons
  in Northern Hemisphere,
    31
  in Southern Hemisphere,
    31
  Sun and, 32
Shadow measurements,
  26–27
Solar system
  asteroids, 60
  planets of, 60
Southern Hemisphere, 31

Stars
  colors, 74
  cycle, 75
  defined, 74
  distances, 75
  temperature, 74
Sun
  path of, 32
  position and sundial,
    40–42
  seasons and, 32
Sundial, 40–42

T

Tides, 13

revolution

solar eclipse

A ___ is
a complete spin on an axis.

A ___ is
one complete trip of one object
around another object

A ___ is a situation that
occurs when Earth, the Sun, and the Moon are in
a straight line and Earth's shadow falls across the
Moon.

A ___ is a blocking of
the Sun's light that happens when Earth passes
through the Moon's shadow.

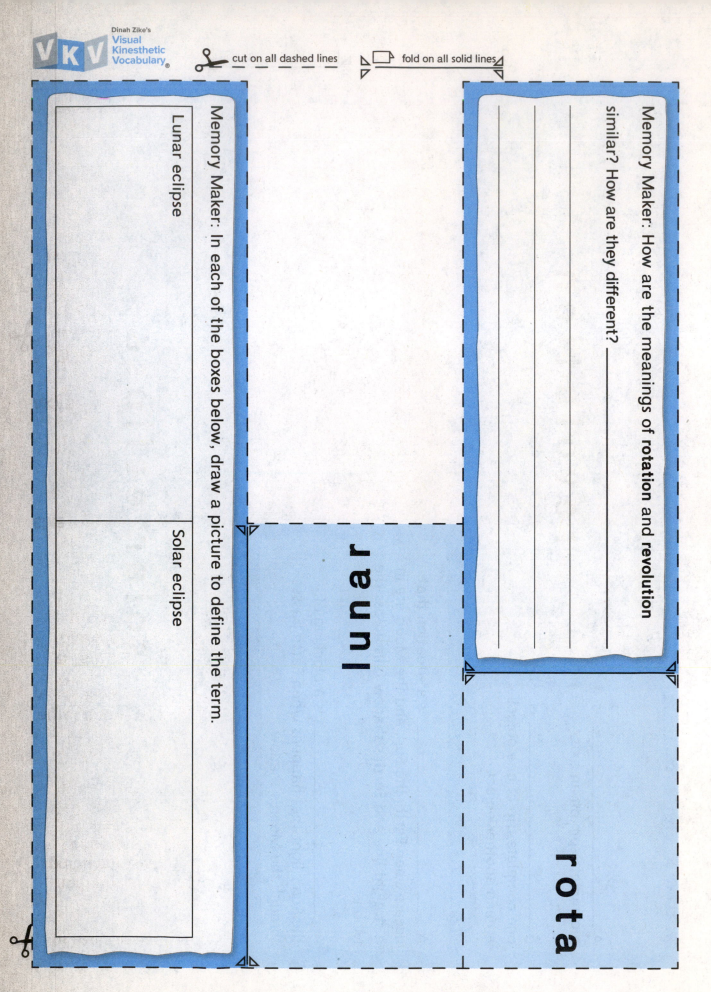

Memory Maker: How are the meanings of **rotation** and **revolution** similar? How are they different? _____

Memory Maker: In each of the boxes below, draw a picture to define the term.

Lunar eclipse

Solar eclipse

lunar

rota

VKV
Visual Kinesthetic Vocabulary®

✂ cut on all dashed lines   📄 fold on all solid lines

constellation

A _____ is
any of the patterns of stars that
can be seen in the night sky
here on Earth.

_____ is the distance light
travels in one year.

A _____ is a measure of the
passing of time as Earth completes one revolution
around the Sun.

year

cut on all dashed lines   fold on all solid lines

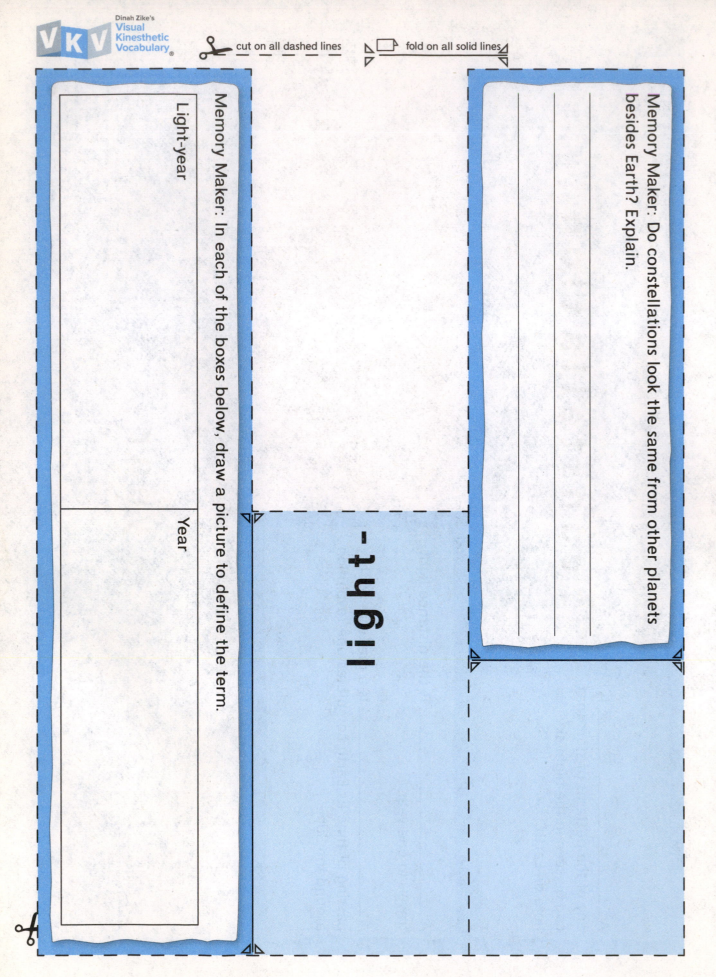

Memory Maker: Do constellations look the same from other planets besides Earth? Explain.

Memory Maker: In each of the boxes below, draw a picture to define the term.

Light-year

Year

light-

VKV4   Module: Earth and Space